D1084872

THE NATURE AND ORIGINS
OF SCIENTISM

PUBLISHED AQUINAS
LECTURES

St. Thomas and the Life of Learning (1937) by the Rev. John F. McCormick, S.J., former professor of philosophy at Loyola University.

St. Thomas and the Gentiles (1938) by Mortimer J. Adler, Ph.D., associate professor of the philosophy of Law, University of Chicago.

St. Thomas and the Greeks (1939) by Anton C. Pegis, Ph.D., associate professor of philosophy, Fordham University.

The Nature and Functions of Authority (1940) by Yves Simon, Ph.D., associate professor of philosophy, University of Notre Dame.

St. Thomas and Analogy (1941) by the Rev. Gerald B. Phelan, Ph.D., president of the Pontifical Institute of Mediaeval Studies, University of Toronto.

St. Thomas and the Problem of Evil (1942) by Jacques Maritain, Ph.D., professor of philosophy, Institute of Mediaeval Studies, University of Toronto.

Humanism and Theology (1943) by Werner Jaeger, Ph.D., Litt.D., "university" professor, Harvard University.

The Nature and Origins of Scientism (1944) by the Rev. John Wellmuth, S.J., chairman of the Department of Philosophy, Loyola University.

The Aquinas Lecture, 1944

THE NATURE AND ORIGINS OF SCIENTISM

Under the Auspices of the Aristotelian Society
of Marquette University

BY

JOHN WELLMUTH, S.J.
Chairman of the Department of Philosophy
Loyola University

MARQUETTE UNIVERSITY PRESS
MILWAUKEE
1944

COPYRIGHT, 1944
BY THE ARISTOTELIAN SOCIETY
OF MARQUETTE UNIVERSITY

B
765
.T5A6
1944

PRINTED AT THE MARQUETTE UNIVERSITY PRESS
MILWAUKEE, WISCONSIN

MARQUETTE UNIVERSITY PRESS
MILWAUKEE

Nihil Obstat

Gerard Smith, S.J., censor deputatus
Milwaukiae, Die 23 Augustus, 1944

Imprimatur

Milwaukiae, Die 23 Augustus, 1944
+ Moyses E. Kiley
Archiepiscopus Milwaukiensis

Imprimi Potest

Die 5 mensis Julii anni 1944
Leo D. Sullivan, S.J.
Praepositus Provincialis
Provinciae Chicagiensis

THE AQUINAS LECTURES

The Aristotelian Society of Marquette University each year invites a scholar to speak on the philosophy of St. Thomas Aquinas. These lectures have come to be called the Aquinas Lectures and are customarily delivered on the Sunday nearest March seven, the feast day of the Society's patron saint.

This year the Aristotelian Society has the pleasure of recording the lecture of the Rev. John Wellmuth, chairman of the department of philosophy at Loyola University. Fr. Wellmuth's studies include A.B., A.M. St. Louis University; Ph.D. University of Michigan; Graduate studies, Oxford University and Theological studies, Weston College.

He is a member of the American Catholic Philosophical Association, American Philosophical Association, Association for Symbolic Logic, and Michigan Academy of Arts, Sciences and Letters.

Fr. Wellmuth is co-founder and editor of the *Modern Schoolman* and has contributed to numerous philosophical publications such as *The New Scholasticism, Thought, America* and the *Loyola Education Digest*. To this list the Aristotelian Society has the honor of adding *The Nature and Origins of Scientism*.

The Nature and Origins
of Scientism[*]

THE title of this lecture manifestly raises
two questions. First, what is scientism?
Secondly, when and how did it begin? In saying what is meant by the word "scientism" we
shall be giving a direct answer to the first
question; but in order to justify a discussion
of this topic we must further show that there
is such a thing and that it is worth considering. The question of the origins of scientism,
like most questions about origins, is not fully
answerable in a single lecture. What we shall
try to do, by way of partial answer, is to outline a few of the causes of scientism, in order
that its nature may be better understood and
its implications more fully appreciated.

The word "scientism," as used in this lecture, is to be understood as meaning the belief
that science, in the modern sense of that term,
and the scientific method as described by modern scientists, afford the only reliable natural
means of acquiring such knowledge as may

be available about whatever is real. This belief includes several characteristic features.
In the first place, the fields of the various
sciences, including such borderline or over-
lapping sciences as mathematical physics, bio-
chemistry, physiochemistry and mathematical
logic, are taken to be coextensive, at least in
principle, with the entire field of available
knowledge. Each of these sciences investigates
and describes a particular kind of reality or
inquires into some phase of reality from a
particular point of view, and the sum total of
their correlated findings represents all that
we know at a given time. This knowledge is
not what used to be meant by "scientific
knowledge," the "scientia" of the ancients.
For the conclusions of science are no longer
regarded as conclusive. More extended researches are likely to lead to further discov-
eries which will not only modify currently
accepted conclusions but may even suggest
new hypotheses to replace those now in use;
so that eventually the picture of the universe

and of ourselves which the sciences present to us today may be even more drastically altered than the scientific view maintained a century ago has been altered by subsequent investigations.

The second characteristic feature of this belief is that the scientific method, as exemplified in the above mentioned sciences, is the only reliable method of widening and deepening our knowledge and of making that knowledge more accurate. Though the different sciences have different techniques and special methods of their own, all agree in making use of the same methodological principles. All begin with careful observation of the phenomena in the field which each has marked out for investigation, and proceed to form hypotheses on a basis of the data so observed. Once these hypotheses have been so formulated as to be adequate to the phenomena in question, and no more complex than need be, a more extended series of observations is conducted, supplemented by experiments if possible, by way of verifying or modifying the initial hypotheses and thus establishing them as theories. It is understood, of course, that this procedure can never give rise to genuine certainty. Progressive verifications of a theory render that theory highly probable, but cannot make it certain. Hence the phrase "theoretical knowledge," which used to mean "certain knowledge," is now taken to mean "hypothetical knowledge," of which certainty is the unattainable limit.

The third characteristic feature of scientism is a more or less definite view about the status of philosophy in relation to the other sciences, insofar as philosophy is itself considered to be a science. This view is likely to find expression in one of two forms: either that philosophy should be made scientific by conforming to the methods and ideals of some particular science, or that the function of philosophy is to correlate and if possible unify the findings of the other sciences by means of generalizing on a basis of these findings, after having rid itself of outworn metaphysical notions. Any system of philosophy which clings to metaphysical principles will simply be superseded by modern science, because such principles are as useless as the inadequate data of the pre-scientific age during which they were arrived at by inductive generalization.

As soon as one understands the characteristic features of the attitude which I have called "scientism," it will be seen at once that this attitude is an actual reality. Because of the development of new sciences within the past hundred years, and especially because of the comparatively recent admission that scientific knowledge does not mean certainty, the scientism of today will not be the same in all respects as the scientism of former times. What is common to scientism in every age is the third characteristic feature mentioned above: the belief that philosophy should be brought into conformity with one or more of the other sciences as regards its ideal or its method.

When Descartes, for instance, attempted to unify the various sciences of his day by giving to them all a mathematical interpretation and demonstrating their conclusions mathematically,[1] the mathematical ideal which he aimed to achieve in philosophy was very different from that of modern mathematics,[2] yet he insisted that philosophical knowledge should emulate a mathematical ideal. Again, the physics of Kant's time was extremely unlike modern physics in many ways; nevertheless Kant's deliberate endeavor to adopt in metaphysics the methods of Newtonian physics[3] was an attempt to make physics the model of philosophy. And Auguste Comte's *System of Positive Philosophy*, according to which the theological and metaphysical states of human culture have been inevitably superseded by the positive or scientific state,[4] is an outstanding example of scientism in spite of the fact that Comte's

ideas about science would be no more acceptable to a modern expert than they were to
John Stuart Mill during Comte's own lifetime.[5]

Scientism as a contemporary phenomenon
is unmistakably recognizable in many of the
books written by scientific experts on philosophical problems. Sir James Jeans, for example, in his recent work entitled *Physics and
Philosophy* (1943), which he speaks of as
"the reflections of a physicist on some of the
problems of philosophy,"[6] is at pains to
indicate the differences between the two sub-
jects of which he is writing, yet it is clear that
for him the new physics has a great deal to do
with such philosophical questions as the nature of causality and the freedom of the
will.[7] The "new philosophy"[8] advocated by
the eminent mathematician, Bertrand Russell, is actually a kind of super science: "It
aims," he tells us, "only at clarifying the fundamental ideas of the sciences, and synthesiz-
ing the different sciences in a single compre-

hensive view of that fragment of the world that science has succeeded in exploring."[9]

That modern philosophers are by no means immune from scientism is equally obvious. Everett W. Hall, of the University of Iowa, has lately presented an excellent account of the status of modern metaphysics in an essay on "Metaphysics" written for the volume entitled *Twentieth Century Philosophy*. As a professional metaphysician, he strongly objects to the "widespread (metaphysical) assumption of our age that the scientific method is the only admissible cognitive method,"[10] and disagrees with the view that "the function of metaphysics is simply to accumulate and perhaps systematize the theories of the sciences." At the same time, he admits that "The task of metaphysics is to generalize further on the basis of scientifically established proportions; i. e., to establish by induction from scientific propositions, propositions that do not occur, at least explicitly, in the sciences at all."[11] Such propositions, it is true, "cannot be the only grounds for meta-physical generalizations."[12] The most im-portant "extra-scientific basis of metaphysics," in addition to critical common sense beliefs, is "direct, sensory experience,"[13] wherein we seek not so much to verify or to disverify our metaphysical generalizations by means of instances, as rather to use instances "illustra-tively, to instigate an imaginative survey," and ultimately we "must rely on a total in-sight."[14] The metaphysical method thus described is "not as reliable as scientific method but more relevant to the sort of hy-potheses involved." In spite of being "highly unreliable," this method is "the best available for metaphysics," but the author warns us that "since the most rigorous scientific method cannot attain certainty, metaphysical proposi-tions should be put forward even more ten-tatively."[15]

This brief statement of Professor Hall's conclusions about the scope and method of metaphysics is not intended as an adverse criticism of his clear and careful essay. The point is simply this: he regards metaphysics as a body of inductive generalizations, none of which can approximate certainty as nearly as do scientific propositions, although they go further than these latter by their reliance on extra-scientific data and their appeal to imaginative insight. Hence in spite of his serious endeavor to avoid the more extreme forms of scientism so well discussed in that essay, he ends by considering metaphysical propositions as less certain than scientific propositions.

It is not the purpose of this lecture to inquire how far the attitude which I have called scientism is justified. An adequate answer to this question could hardly be given without investigating the nature of philosophy and the nature of the positive sciences, and attempting to determine the proper function of their respective methods as instruments of human knowledge. There are two remarks, however, which should be made in this connection. The first is that the investigation above mentioned cannot possibly be undertaken by any or all of the positive sciences, because in that case it would have to be carried on by the scientific method, and hence could never lead to any conclusive results even if some means were found of applying this method to this particular problem: a supposition which, to say the least, is extremely doubtful. The scientific method, as employed in modern science, admittedly rests on a number of assumptions regarding the nature of truth and the criteria for distinguishing between truth and error, as well as on a more or less definite aggregate of useful beliefs which are explicitly referred to nowadays as "the faith of the scientist."[16] Further, the scientific method itself includes the belief that every principle is an assumption or hypothesis, to be accepted only insofar as it can be verified by observation and experiment and insofar as it is suggestive of other hypotheses. The adoption of the scientific method as an instrument of investigation therefore commits one to the position that no certain conclusions will be reached.

It should be remarked, in the second place, that some form of scientism is almost certain to be adopted by our contemporaries in the field of philosophy and the field of science, because they can hardly avoid believing that philosophy is a failure. Every one of the systems of philosophy now in vogue has been more or less directly influenced by the philosophy of Descartes, who is not only the father of modern philosophy but the father of philosophy itself for many of our modern thinkers. When Sir James Jeans writes a book on *Physics and Philosophy,* the only philosophy he considers worth discussing is the philosophy of the past three hundred years; and he remarks, as though there were nothing more to be said on the matter, that neither the philosophical study nor the physical research of this interval "has shown any cause for changing Descartes' dicta" about efficient causality and the nature of free will.[17]

Since Descartes' ideal of certainty in philosophy has vanished with the progressive decline of his system through Locke and Berkeley into the empiricism of Hume, and since neither Kant with his Copernican revolution nor Hegel with his attempt to fuse into organic unity the three levels of the Kantian structure has provided an acceptable substitute, it is only natural that contemporary thinkers should question the validity of metaphysical speculation and should tend to rely on the faith of the scientist. If modern science needs a philosophical background, Descartes' philosophical atomism is uncritically considered better for this purpose than a philosophy of matter and form, especially in the light of the atomic theory of matter.[18] But for the most part, the faith of the scientist rests on the success achieved in both the speculative and the practical order by modern scientific theories.

What I have said thus far about the nature of scientism will surprise no one who is acquainted with the William James Lectures delivered by Professor Gilson at Harvard University in 1937 and published under the title of *The Unity of Philosophical Experience.* The really surprising thing is that this book has made so little impression on contemporary thought; indeed, the fact that it has not received sufficient attention is a striking indication of how widespread scientism is today. It is true that many of our contemporaries have begun to realize the inadequacy of any science, or of all the sciences, to provide answers for their philosophical questions, but they may well feel obliged to accept this inadequate substitute as the only available source of answers, apart from supernatural knowledge. And this situation is likely to continue until they either give up the quest for knowledge or else discover a set of philosophical principles and a method of philosophizing which will be more successful in coping with philosophical problems than the systems of Descartes and his successors.

Now, most of us here believe what some of us already know on a basis of personal experience: that the philosophy of St. Thomas Aquinas contains just such a set of philosophical principles and just such a method of philosophizing. It would be absurd to suggest that he anticipated and even solved all the important problems of our time; but those who have made a first hand study of his philosophy are prepared to maintain that although he was mainly concerned with establishing a Christian philosophy on an Aristotelian basis in order to avoid some of the difficulties inherent in the Platonic tradition, and was especially interested in determining the relations between faith and reason, he was also alive to the problem of the relation between philosophy and the other sciences. Besides carefully delimiting the respective spheres and methods of natural and supernatural knowledge,[19] he established within the field of natural knowledge itself a clearcut distinction between philosophy and the other sciences;[20] and in spite

of the fact that the sciences of his day and
their methods were different from our own,
the principles which he laid down enable us
even now to distinguish the proper method
of philosophy from the modern scientific
method, and to tell the difference between a
scientific problem in the modern sense and a
philosophical problem.[21]

I am not here concerned to examine the
basis of these claims or to evaluate their
soundness. I mention them only because of a
serious difficulty which presents itself as soon
as one hears them made: a difficulty so serious
as to prevent them from receiving the consid-
eration which they may deserve. If the philosophy of St. Thomas, as elaborated by him during the third quarter of the thirteenth century, did in fact have within itself that which
should have made scientism impossible, why
did it not actually prevent scientism from aris-
ing, or at least serve as a barrier against the
spread of scientism? By way of suggesting
how to solve this difficulty, I would ask you

to consider with me, in this second part of my
lecture, the question of the origins of scientism.

The first answer that will occur to anyone
who asks why the philosophy of St. Thomas
did not stop the growth of scientism is that
the positive sciences and the scientific method
proper to them were entirely undeveloped in
his day, waiting for the Renaissance, or even
for Francis Bacon, to release men's minds
from the shackles of theology and authority
and the cult of the four idols. After all, it
will be said, the philosophy of Aquinas was
based on the very inadequate and erroneous
scientific theories of the time, and was more-
over supported by the authority of the Catholic Church. Naturally enough, as soon as the
rise of the positive sciences forced these early
theories into the discard, and especially when
the Reformation freed men from authoritarianism in their thinking, the system of Aquinas
simply fell to pieces as did all the other systems of medieval thought; in which state, inci-

dentally, it would have remained till now, had not a group of Catholics, under the influence of Papal authority, felt bound in conscience to salvage the wreck and construct that strange anachronism which is known today as neoscholastic philosophy or neothomism.

Widely accepted as this answer has been, and deeply entrenched as it is in the minds of our contemporaries,[22] it hardly does justice to historical fact. Of its non-historical aspects nothing can be said here, although one such aspect stands in need of extended treatment, namely: the confused issues raised by such stock phrases as "authoritarianism in thought." It may be noted in passing that the adequacy of this answer is beginning to be questioned by some of those who still adhere to it; for the growing appreciation of what St. Thomas' philosophy really means has caused more disquiet among modern thinkers[23] than would ever be caused by a structure of thought, however imposing, erected

out of ruins into a precarious and wobbly up-
right position. The brief survey of the origins
of scientism which we are about to make is
intended not so much to make plain the inadequacy of the above answer, as rather to suggest that an alternative answer is perhaps
more adequate. That alternative answer is
simply this:

The movement called scientism needed no
Renaissance to give it life, nor any fullgrown
scientific method to foster its development. It
was the natural outcome of a trend of thought
which began in the early Middle Ages, which
was strongly opposed by St. Thomas during
his lifetime, and which ultimately led to the
breakdown of medieval philosophy before the
end of the fourteenth century, at a time when
the traditional fathers of modern science had
still about two centuries to wait before being
born.

This trend of thought may be described,
quite generally, as a gradual loss of confidence in the power of the human mind. The

following attempt to indicate the various stages through which it passed will make clear some of the causes that contributed to it.

As is well known, one of the main problems which concerned the thinkers of the Middle Ages was the problem of the precise relationship between faith and reason.[24] Obviously, this problem is very largely a philosophical one, involving as it does the nature of human thinking, the starting point of human knowledge, the mind's power of arriving at the truth, and the means by which it can grasp the truth with certainty. The connection between this problem and the problem raised by scientism is easy to see: for in both cases there is question of two ways of arriving at truth, and of the various conditions, methods and means proper to each. That there is also a historical connection between the two problems will become clear on examination.

In the middle of the ninth century, Scotus Eriugena insisted on the supremacy of reason

over human authority, even such accepted
authority as the Fathers of the Church.[25]
When Berengarius of Tours took the further
step, two centuries later, of considering human reason superior to the word of God,[26]
there was a reaction against philosophical reasoning which led to the long quarrel between
dialecticians and theologians all during the
eleventh century,[27] until St. Anselm finally
prevailed over the antirationalists of his time
by insisting on strictly philosophical demonstrations[28] and on the utility, not to say
necessity, of philosophical reasoning in order
to understand the truths of faith.[29] His reliance on the power of human thought was
so great that he did not hesitate to propose
proofs from reason of such mysteries as the
Trinity and the Incarnation.[30] The same
emphasis on philosophical proofs which must
rest on the data of experience is to be found
during the next century in Richard of St.
Victor,[31] although that century was more
remarkable for its interest in logic and the

humanities and in mysticism than for the construction of philosophic systems.[32]

The problem of faith and reason was of course not confined to Christian thinkers. For the Arabians it took the form of establishing the proper relationship between philosophical speculation and the doctrines of the Koran,[33] and the Moslem theologians violently took issue with the philosophers, as is clear from Algazel's treatise, *The Destruction of the Philosophers*.[34] It is perhaps worth noting here that the solution offered by Averroes, according to which philosophy is ranked above theology and the lowest place is given to faith and to religion, need not mean an absolute hierarchy of different truths, but may well be a description of three possible attitudes toward one and the same truth. He appears to hold that although Truth itself is to be found only in the Koran, this truth will be grasped in different ways and in different degrees by different minds. The true philosopher will not be content with anything short of strict demonstration; the theologian, a mere dialectician, will rest satisfied with probable arguments; while the ordinary believer will assent on a basis of imaginative or rhetorical arguments.[35]

Even if the Jewish philosophers had not come under Arabian influence, they would have had to face a similar problem in attempting to reconcile the conclusions of philosophy with the teachings contained in their Bible. With them, as with the Christians and the Arabians, we find a strong theological reaction against philosophy and philosophers; indeed the outstanding representative of this reaction, Juda Hallevi, seems to have been more sincere in his convictions than the Arabian antiphilosopher Algazel,[36] though less vigorous in his condemnation of earthly learning than was St. Peter Damian a generation earlier.[37] Their nearest approach to a solution of the problem, as found in Moses Maimonides' *Guide to the Perplexed*, consisted in regarding philosophy as a kind of apologetic, its proper object being to afford a rational confirmation of the Law.[38] At the same time, Maimonides insisted on the importance of philosophical speculation by stressing man's duty to practice philosophic thinking during this life as a means of enriching his mind.[39]

Apart from other noticeable similarities, the common element in all these positions appears to be this: the conviction that there can be no contradiction between the truth as known by reason and the truth as made known through the word of God which we believe by faith, since God is the source of all truth and of all reality. The divergencies in each of these views are due to a lack of agreement about the precise relationship between philosophical thinking and religious belief. Extremists maintain on the one hand that no revealed truth is credible unless demonstrable by reason, and on the other hand that human reasoning is vain and that belief should take the place of philosophical thinking. Midway

between these extremes we find some difference about how far human reason can go in
demonstrating the truths of faith; but the tendency is to be quite optimistic on this point, at
least among Christian thinkers who hold that
faith is a needed preliminary for understanding.[40]

This problem was still very much alive in
the time of St. Thomas. Whether it was definitively solved by him, or by his contemporary
St. Bonaventure, or by both, is a question
which need not be answered here. The point
to notice is that in solving the problem, each
of these two thinkers put forward a well defined theory of knowledge, the consequences
of which in each case became apparent long
after the death of its author.

According to St. Bonaventure, who followed in the main the tradition derived from
Plato by Augustine, through Plotinus all our
knowledge of intelligible things is due to a
divine illumination which is a kind of created
reflection of the divine ideas whereby God

knows things.[41] Contingent and sensible knowledge can indeed be had without this divine light, but such knowledge is not certain except in a restricted sense: for things that are beneath us can give us only relative certainty, whereas absolute certainty comes only from higher things.[42]

St. Thomas, whose theory of knowledge was not Augustinian but Aristotelian, had no such Platonic misgivings about the comparative unintelligibility of the world of sensible objects. While maintaining the dependence of the human intellect on God in its every act of knowing, he insisted that in virtue of its God-given nature, without the need of special light from on high, the mind is able to know things with certainty through the intelligible forms which it can discern in them through its power of abstraction; and that its knowledge is derived from things themselves by this power which is the light of the agent intellect, rather than by a kind of reflection within us of the divine ideas of things.[43]

The significance of these two theories can be appreciated only when we realize the spirit which animated each of them. Whereas St. Thomas was careful to establish the certainty of our knowledge of material reality on a sound philosophic basis by his analysis of the nature of physical objects[44] and the nature and functions of the knowing mind, St. Bonaventure was less interested in determining what these objects were than in discovering what they signified, as created expressions of God's attributes.[45] Not only did he consider a knowledge of the nature of these things relatively unimportant, but he attributed all certain knowledge to a divine illumination rather than to the mind's exercise of its natural power of knowing.

The consequences inherent in this latter view can hardly have been evident to St. Bonaventure himself, but as we shall see, they were made clear enough by those who followed him. We cannot study in detail the various events which had a bearing on the rapid emergence of these consequences and on the almost complete neglect of the view of St. Thomas. In a different historical setting, the Aristotelianism of St. Thomas might have exerted a more powerful influence on the thought of his time, instead of being identified to some extent with the intransigent Aristotelianism of the Latin Averroists.[46] Such contingencies as this, extrinsic to his system of philosophy, may help to explain why the oft-mentioned triumph of Thomism was rather potential than actual during the centuries after his death, and why the history of philosophy went on almost as though St. Thomas had never existed.[47] In a different historical setting, also, the consequences implicit in the philosophy of St. Bonaventure might never have been developed. But their importance for the history of philosophy consists in the fact that they were there to be developed; and hence their actual development, though occasioned by contingent historical events, was at the same time a natural out-

growth of the system which contained them
in principle.

The first stage in this development is ob-
servable among the pupils of St. Bonaventure,
who carried to extremes the depreciation of
our knowledge of sensible things. The difference between such knowledge and strictly
certain knowledge was accentuated by
Matthew of Aquasparta, who held that two
kinds of intelligible species are necessary, one
whereby we have universal knowledge.[48]
It was apparently to counteract this view that
Richard of Middleton introduced his distinction between the psychological universal and
the logical universal, the former being an actual existent singular which exists not mate-
rially but in the mind, "with an existence
more real than if it existed in some physical
object," and the latter having no real existence
at all, either in the mind or outside, but only
"a represented being, which, however, is
enough to move the mind to act through the
medium of the above mentioned species," i.e.,

through the psychological universal.[49] One
can easily understand how such a "represented
being" would fall a ready victim later on to
Ockham's razor, leaving only the psychological universal, with the result that such
terms as "man," "animal," "living being,"
would stand only for concepts or psychological entities, not for some real extramental
nature shared by all existing individuals of a
species or genus. Incidentally, Meinong's
"Objektive," which is very like Richard of
Middleton's "esse representatum," was rejected by recent logicians on much the same
grounds and by the use of the same razor.[50]

Duns Scotus, while following the philosophy of St. Bonaventure, apparently agrees
with the fundamental theses of St. Thomas;
yet some of these theses underwent at his
hands some changes that were far reaching in
their consequences. Whereas St. Thomas
holds, with Aristotle, that *a posteriori* demonstrations are real demonstrations, though in-
ferior to those which are *a priori,* Scotus con-

siders them to be demonstrations only in a modified sense: "No demonstration which goes from effect to cause is demonstration in an unqualified sense."[51] At this stage there emerges a very significant element of the trend of thought we are following: the field of philosophical demonstration is beginning to be narrowed down, and the result is at once manifest. For Duns Scotus declares that some attributes of God, especially His providence over rational creatures, are merely "believable": that is, they can be known with certainty by faith alone; and as for the other attributes, such as First Cause and Ultimate End, which were formerly thought demonstrable *a posteriori*, "by natural reason we can come to some sort of conclusions about them."[52] In spite of the difficulties which troubled his unusually acute and critical mind, Duns Scotus himself had no serious doubts of the validity of metaphysics as a speculative science.[53] But when lesser minds were confronted with these same difficulties they

adopted an over-critical attitude. This attitude was reflected in John of Polliaco's re-
jection of the proofs of God's infinity because of their dependence on the fact of cre-
ation, which itself could not be proved,[54]
and in Peter Aureolus' denial that the soul
could be proved to be the form of the
body.[55] It is easy to see that a whole important body of philosophical conclusions
was thus gradually shifted to the field of
theology, so that their certainty was a matter
of faith and not of reason. What needs to be
pointed out, because it is less obvious, is the
fact that theology itself thus ceased to be a
speculative science, and became a practical science, the function of which was to direct
men's actions rather than to increase or deepen
their knowledge. For rationally indemon-
strable propositions are not knowable with
certainty, and hence are not objects of scientific knowledge in the strict sense.[56]

Scotus' well known distinction between
intuitive knowledge and abstractive knowl-

edge, which seems to have been connected
with the illumination theory of knowledge as
proposed by St. Bonaventure,[57] is important
less for what it is than for what was later
done with it. On both the sense level and
the level of intellect, he calls "intuitive" that
knowledge whereby we know things that are
actually existent and present, whereas "abstractive" knowledge is the intellectual knowledge which we have of essences, abstracting
from their existence, and the sense knowledge
which we have of things not present but
imagined.[58] What use Ockham made of this
distinction, we shall see shortly.

But before considering Ockham's philosophy and its place in the trend of thought
which we are following, there are two other
factors that contributed to this trend which
must be mentioned here. The first has no connection whatever with the principles of St.
Bonaventure's theory of knowledge, but is
the outgrowth of the vigorous debates between opposing schools of thought at both

Paris and Oxford during the fourteenth century. In this connection the influence of the *Summae Logicales* is especially notable. This debater's manual, compiled in the thirteenth century by Peter the Spaniard from an earlier *Summa* of Lambert of Auxerre, was widely used by those who took part in the University "disputations." It is not really a manual of logic, though Buridan's later additions made it appear so; for it was originally based on the *Topics* rather than on the other treatises of the *Organon,* and the discussions in it are explicitly said to be merely probable, with no reference to demonstration and exact knowledge.[59]

By the many distinctions and subdistinctions which Peter the Spaniard introduced into the notion of "term," he paved the way for the later "terminists" of Ockham's time,[60] and his emphasis on probability had the following disastrous effect. Propositions which were admitted by all as certainly true were put forward as only probable when used as objections in the disputations, and on the other hand, admittedly false propositions were treated as probably true because arguments, however specious, could be proposed in their defense. Now this practice of dialectic, which had some value for developing skill in argument, would of itself have been no more productive of skepticism than the practice nowadays of arguing for and against some political or economic measure in high school and college debates is productive of an immediate change in our political or economic system. But in conjunction with the tendency already noted, to doubt the validity of philosophical demonstrations, this procedure was bound to have an unsettling effect: especially since it permitted men to air all sorts of views as probable, without actually sponsoring these views as their own. A typical illustration is the following statement of Peter of Candia: "This is the stand which I took on my second principle against Master G. Calcar; and now I am maintaining the opposite here, not because I think one is more true than the other, but to bring in some various shades of colorful thinking."[61] The same note was sounded in discussing important problems of philosophy, especially in the debates on free will at Oxford. Thus Thomas Buckingham, speaking of God's influence on the will, begins his remarks as follows: "As regards the main topic, without assenting to this at all but nevertheless for the sake of practice in debating, I say that God wills sin to occur and to exist, and wills man to sin mortally and venially."[62]

The second factor which gave freer scope to the later ideas of Ockham was the cultivation of a type of philosophy at Oxford which was rather scientific than metaphysical. It began with Robert Grosseteste, the teacher of Roger Bacon, who was markedly influenced by the Arabian treatises on optics and mathematics.[63] To say that his "metaphysics of light" is an unconscious attempt to substitute physics for metaphysics may be an exaggeration. But it remains true that by taking light to be the first form created by God in prime matter,[64] he does seem to be identifying a physical reality with a metaphysical principle, thus blurring the sharp distinction between "something that exists" and "that by which something else exists," a distinction on which the whole theory of matter and form depends. Roger Bacon was inspired with the thought of making science the handmaid of theology,[65] but in order to do this he felt it necessary to get rid of philosophy as an independent discipline, by referring to the illumination theory of knowledge which he interpreted as a kind of revelation. "The reason," he says, "why philosophical wisdom is reducible to theological wisdom is not only that God has illumined men's minds for acquiring a knowledge of wisdom, but that they possess wisdom itself from Him, and He has revealed it to them."[66]

Having thus converted philosophy into theology, he left the field of human knowledge wide open for the sciences. Impressed

by the use which Robert Grosseteste and
Adam Marsh had made of mathematics,[67]
he stated very forcefully, "It is impossible for
the things of this world to be known without
a knowledge of mathematics,"[68] and as Peter
of Maricourt had done before him, he insisted on the need of sense experience to
make mathematical evidence itself more convincing.[69] The most perfect knowledge of
all, according to him, except for that purely
interior and spiritual knowledge which culmi-
nates in mysticism, is experimental knowledge,
i. e., knowledge based on sense experience directly: for it alone provides a complete and
detailed demonstration of the conclusions of
other sciences, and can demonstrate truths
which they cannot prove by their own proper
means, besides being able to discover the se-
crets of nature and so giving power to those
who possess it.[70]

Both these factors were prominent in the
immediate background of William of Ockham. The way was prepared for his particular

type of conceptualism by the earlier views of
some of the Paris theologians and masters of
arts. One of the latter, John of Polliaco, whose
opposition to Dominicans and Franciscans is
very like that of William of St. Amour, denied the need of any intelligible species for
knowledge. The Dominican Hervaeus Natalis
took over Scotus' distinction between intuitive
and abstractive knowledge, and held like
Richard of Middleton that universals have
only a mental existence (esse objectivum).
Another Dominican, Durandus of St. Pour-
cain, emphasized the same notion in the first
edition of his *Commentary on the Sentences*.
Speaking of universals such as genera and
species, he insists that they have no reality
whatever either outside the mind or in the
mind, though they are in the mind "objec-
tively." Just as these two Dominicans were
dissatisfied with the philosophy of St. Thomas,
so the Paris Franciscan Peter Aureolus was
out of sympathy with the philosophy of Duns
Scotus and indeed with all other systems of

his time. Influenced partly by Averroism and partly by Durandus, he denied the extramental reality of universals and worked out a theory of knowledge according to which abstract knowledge is merely a faded copy of that knowledge of individuals which alone puts us in touch with reality.[71]

William of Ockham simply carried these notions to their logical conclusion. According to Scotus, it will be remembered, the distinction between intuitive and abstractive knowledge is applicable both to intellect and to sense. Intuitive sense knowledge, which Ockham calls "experimental," is the direct result of the object's acting on the sense organs, and so stimulating perception of sensible properties; whereas abstractive sense knowledge is the representation of absent objects by the imagination. Intuitive intellectual knowledge, which Ockham considers not the same as judgment, is based on intuitive sense knowledge, and according to him, makes us aware of the existence of an object while the object is acting on our senses. Abstractive intellectual knowledge, for Ockham, is the intellectual representation of a formerly perceived existent: it is therefore a concept which abstracts from existence.[72] Since he holds that from abstractive knowledge nothing can be concluded about the existence or non-existence of the object of such knowledge, it is by intuitive knowledge alone that we perceive the existence or non-existence of things. To account for an intuitive knowledge of their non-existence, he was obliged, if he wished not to modify his entire position, to hold that God produces in us all the intuitions which we have of existing objects, and can therefore also produce in us an assent, even when the object does not exist, very like the evident assent which accompanies intuitive knowledge.[73]

In the reaction which followed Ockham's denial of universals, two men made a noteworthy attempt to rescue philosophical knowledge from skepticism. John of Mirecourt, go-ing back to St. Augustine, based all knowledge on the direct experience which we have of our soul, which he maintains is the only substance that we know with certainty. Natural knowledge is of two kinds: knowledge of the first degree, which includes analytic judgments and their consequences, as well as knowledge of one's own existence and of subjective states. This knowledge is absolutely certain, and is made to depend in part on a principle which echoes St. Augustine: "What doubts must exist." The second kind, which he calls knowledge of the second degree, or experimental knowledge, is directly connected with Ockham's intuitive knowledge. Through it we have knowledge of the external world which is certain and evident but not infallible.[74]

Not satisfied with this, Nicholas of Autrecourt attempted to limit the field of certainty as narrowly as possible so as to make at least this narrow area secure.[75] According to him, there is no certainty except immediate evidence, which comes only from two sources: experimental knowledge, and the principle of identity, with the cognate principle that contradictories cannot be true together. All knowledge based on this principle is certain; hence every true syllogistic conclusion must be reducible to it, and in all such arguments the antecedent and the consequent must be wholly or partially identical: for if they are not, the contradictory of the consequent may be asserted without contradiction. One consequence of this view is that the principle of causality is worthless, as is also every proof based on it. The bond between effect and cause is indeed experimentally evident, but this is no guarantee of its necessity, and we cannot prove from the existence of an effect that the cause of this effect must exist. The difficulty is that although experience gives evidence of a fact, we have no more than probability that the fact will occur again.[76] Roger Bacon had said, in extolling the superiority of experimental knowledge, that only experience of combustion would give convincing evidence of the truth that fire burns.[77] Nicholas of Autrecourt is much less sanguine. "Because," he says, "it was once evident to me that when I placed my hand near a fire I was warm, it is therefore probable to me that I should be warm if I were now to place my hand there."[78] We have direct knowledge of our own souls, he admits, and therefore of their existence; but we cannot affirm even with probability that substances exist, for we can have no direct experience of substances.[79] When addressing his fellow scientists of the Royal Society of Edinburgh on October 26, 1942, Professor E. T. Whittaker of the University of Edinburgh referred to the way in which the virile scholars of the Renaissance broke away from the futile subtleties of the degenerate Schoolmen, and told his hearers how "at Paris, in 1536, a crowded audience proclaimed the thesis of Peter Ramus, 'Whatever is in Aristotle is false'."[80] It is clear that Professor Whittaker does not regard the earlier Schoolmen as degenerate and preoccupied with subtleties; but the virile scholars of the Renaissance were not particularly original in their breaking away. For just two hundred years earlier, Nicholas of Autrecourt was saying that "in the whole of the natural philosophy and in all the metaphysics of Aristotle, there are not two conclusions that are certain, and perhaps not even a single one."[81]

In this atmosphere of complete lack of confidence in human knowledge, the "new physics" developed by the later Ockhamists and others marked the first beginnings of modern science.[82] There is no need of outlining that development here: Buridan's analysis of movement, based on the idea of impetus which Peter Olivi had conceived forty years earlier;[83] the abandonment of Aristotle's notions about the heavens; and the growing belief in the rotation of the earth, which had never quite been lost sight of since Scotus Eriugena, though it was still considered a

matter of belief rather than of demonstration
by Nicholas of Oresmes because he was
aware of the principle of the relativity of
movement.[84] What matters for us is the
fact that this development took place in circumstances
most favorable to the rise and
growth of scientism. A brief review of the
data already considered will make this clear.

We have observed in the first place how
the tendency to regard sensible objects from
the standpoint of their symbolic significance
rather than from the standpoint of their complete
reality as created natures, coupled with
the idea that these objects are unknowable
apart from a divine illumination, prepared the
way for a gradual undermining of the validity
of intellectual knowledge which ended in
nominalism. At the same time, and as a natural
consequence, insistence was laid on the
relative superiority of knowledge gained by
direct sense experience, which was needed to
confirm even mathematical knowledge. Next,
the certainty of philosophical propositions

was rendered suspect by the use of probable arguments in debate, until the very possibility of demonstration began to be doubted. As men lost confidence in philosophy, they transferred as many problems as possible to the field of theology, to be accepted on faith: the immortality of the soul, the existence of free will, and eventually the very existence of God and the question whether there is one God or many. What could not be accepted on faith remained uncertain, for at best the arguments in favor of these points were simply more probable than the opposite arguments: and having rejected the validity of notions like substances and accident, they ended by denying the principle of causality itself. Finally, that sense knowledge which had been considered superior to intellectual knowledge was no longer held to be infallible, and no amount of repeated experiences could give more than probable grounds for belief in natural phenomena. Thus, before the end of the fourteenth century, though the positive sciences were only beginning to develop, we have the essential features of the modern scientific method with its emphasis on probability as the ideal of scientific knowledge, and the essential characteristic of scientism at least in this negative sense, that the whole field of human knowledge, apart from revealed truth and theology, was to be explored by other than philosophic means because philosophy had failed.

How successful later philosophies were in redeeming this failure, the first part of this lecture has tried to suggest. In conclusion I should like to remark that there is one philosophy which did not fail, for it was never really tried; and that is the philosophy which is the main topic of this series of Aquinas Lectures.

NOTES

*PREFATORY NOTE: My indebtedness to the researches of Professor E. Gilson and of the Abbé K. Michalski must be acknowledged at once, lest the reader suppose that this lecture is based primarily on an independent study of the works of the later medieval philosophers mentioned in these pages. For the most part, I have been content to examine in their contexts the texts quoted by these two scholars, and have therefore referred to their works rather than to the original sources, some of which, extant only in manuscript, are at present inaccessible. The full titles of the works most frequently cited in these notes are as follows:

The Unity of Philosophical Experience, by E. Gilson (New York, 1937).

La philosophie au moyen âge, by E. Gilson (Paris, 1930).

Les courants philosophiques à Oxford et à Paris pendant le XIVe siècle, by Abbé K. Michalski. In *Bulletin International de l' Académie Polonaise des Sciences et des Lettres* (Cracow, 1919-1920, pp. 59-88).

Les courants critiques et sceptiques dans la philosophie du XIVe siècle, by Abbé K. Michalski. In *Bulletin* (as above), (Cracow, 1925, pp. 41-122).

La physique nouvelles et les différents courants philosophiques au XIVe siècle, by Abbé K. Michalski. In *Bulletin* (as above), (Cracow, 1927, pp. 93-164).

Friedrich Ueberweg's *Grundriss der Geschichte der Philosophie,* revised by Dr. Bernhard Geyer (11th

Edition, Berlin, 1928). All citations are from Vol. II, Die Patristische und Scholastische Philosophie. (Referred to here as "Ueberweg.")

1.   Cf. E. Gilson, *The Unity of Philosophical Experience* (New York, 1937), Chapter V, pp. 125-151.

2.   A short account of the three main schools of thought among modern mathematicians is given in Max Black's *The Nature of Mathematics* (London and New York, 1933).

3.   "The true method of metaphysics is fundamentally the same as that which Newton has introduced into natural science, and which has there yielded such fruitful results." Quoted by Gilson, *op. cit.* p. 227, from Kant's so-called "prize essay" of 1763. Gilson's entire Chapter IX is relevant here.

4.   Cf. *Ibid.*, Chapter X, especially pp. 251-253.

5.   Cf. J. S. Mill, *The Positive Philosophy of Auguste Comte* (New York, 1875), pp. 154-160, 181-182.

6.   *Op. cit.*, Preface, p. vii.

7.   E. g., p. 215: "Before the era of modern physics, it was a simple matter to define what we meant by causality and free will. . . . But modern physics shows that these formulations of the questions have become meaningless."

8, 9.   Cf. Russell's essay, "Philosophy of the Twentieth Century", in *Twentieth Century Philosophy*, edited by Dagobert Runes (New York, 1943). The passages quoted are on pp. 241 and 249.

10.   *Op. cit.,* p. 169.

11.   *Ibid.,* p. 181.

12.   *Ibid.,* p. 182.

13.   *Ibid.,* p. 185.

14.   *Ibid.,* p. 190.

15.   *Ibid.,* p. 191.

16.   This phrase is used as the title of section 1, Chapter XXI, in L. S. Stebbing's *A Modern Introduction to Logic* (London, 1930). The same idea is expressed by Sir Arthur Eddington in the 1938 Tarner Lectures. See his *The Philosophy of Physical Science* (Cambridge, 1939) ; e. g., p. 222: "In the age of reason, faith yet remains supreme, for reason is one of the articles of faith."

17.   Sir James Jeans, *Physics and Philosophy* (Cambridge and New York, 1943), p. 214. His brief comments on medieval philosophy (pp. 18-19) are as inaccurate as they are inadequate.

18.   A careful study of this entire problem in its historical setting is to be found in P. Hoenen, S.J., *Cosmologia* (Rome, 1936) pp. 136 ff.

19.   Cf. the texts referred to in Gilson's *Reason and Revelation in the Middle Ages* (New York, 1938).

20.   Of special importance is the commentary *In Boethii de Trinitate,* QQ. IV to VI.

21.   Perhaps the best known recent use of these principles is that made by M. Maritain in Part I of *The Degrees of Knowledge* (New York, 1938), pp. 27-85.

22.   The extent and influence of this view is difficult to estimate at present because no modern scholar could so completely ignore recent contributions to the study of medieval philosophy as to express the above ideas in a serious article or book. Current discussions such as the one referred to in the following note suggest the prevalence of these ideas outside of professional philosophical circles.

23.   Cf., for instance, the extended series of articles under the general title, "The New Failure of Nerve," in *The Partisan Review,* January-February and March-April, 1943.

24.   Cf. Gilson's book, note 19 above.

25.   E. Gilson, *La philosophie au moyen âge* (Paris, 1930), pp. 13-15. Also the texts quoted in Ueberweg, p. 170.

26.   *La philosophie au moyen âge,* pp. 33-35.

27. *Ibid.,* pp. 35-38. Texts in Ueberweg, pp. 185-6.

28. *Ibid.,* p. 47.

29. *Ibid.,* p. 43, quoting *Cur Deus homo,* c. 2: "Just as right order requires us first to believe the deep matters of faith before venturing to examine them carefully by reason, so I consider it negligence, once we have been solidly established in faith, if we do not study to understand what we believe."

30. He explicitly says that his *Monologion* and *Proslogion* were written so that "what we hold according to faith about the divine nature and persons, apart from the Incarnation, might be proved by necessary arguments without Scriptural authority"; and in *Cur Deus homo* he establishes the necessity of the Incarnation by similar arguments. Cf. texts in Ueberweg, p. 194-195.

31. *La philosophie au moyen âge,* p. 87; texts in Ueberweg, pp. 267-268.

32. *Ibid.,* pp. 91-93.

33. *Ibid.,* pp. 98-99.

34. *The Unity of Philosophical Experience,* pp. 34-38. See also his article, "Pourquoi saint Thomas a critiqué saint Augustin," in *Archives d'Histoire Doctrinale et Littéraire du Moyen Age,* Vol. I, 1926.

35. *La philosophie au moyen âge,* pp. 105-107.

36. *Ibid.,* 112-113.

37. *Ibid.,* p. 37.

38, 39. *Ibid.,* pp. 113-115.

40. This latter tendency was due largely to St. Anselm.

41. *Ibid.,* pp. 153 ff. Cf. also his *The Philosophy of St. Bonaventure* (New York, 1938).

42. He says of the human mind, "ab inferiori recipit certitudinem secundum quid, a superiori vero recipit certitudinem simpliciter" (quoted in Gilson, *La philosophie au moyen âge,* p. 154).

43. For an outline of St. Thomas' position, see chapters XII and XIII of Gilson's *The Philosophy of St. Thomas* (St. Louis and London, 1939).

44. Perhaps the fullest study of this problem is Aimé Forest's *La structure métaphysique du concret selon saint Thomas d'Aquin* (Paris, 1931).

45. This position, mentioned by Gilson, *La philosophie au moyen âge,* pp. 145, 148-149, is more fully outlined in *The Unity of Philosophical Experience,* pp. 49-55.

46. *La philosophie au moyen âge,* pp. 202-203.

47. Forest, *op. cit.,* p. 1, Introduction.

48. Abbé K. Michalski, *Les courants philosophiques Oxford et à Paris pendant leXIVe siècle*. In *Bulletin International de l'Académie Polonaise des Sciences et des Lettres* (Cracow, 1919-1920, pp. 59-88), p. 60.

49. "Quamvis autem universale non existat, tamen eius species realiter in intellectu existit: quia, quamvis non existat materialiter, tamen cum sit res aliqua et existens in re vera, realem existentiam habet et realiorem quam si existeret in aliquo subjecto corporali . . . Et etiam quod est universale, quamvis realem existentiam non habet, sub ratione qua universale, tamen habet esse repraesentatum, quod esse sufficit ad movendum intellectum mediante praedicta specie." Quoted by Michalski, *ibid.*, from Richard of Middleton's commentary *In II Sent.*, *d. 3, princ. 3, q. 1, fol. 17-20* (ed. Venet. 1509).

50. The modern counterpart of the "esse repraesentatum" is what logicians nowadays call "a proposition". Disputes about the reality of propositions in this sense were carried on at intervals in *Mind* and in the *Proceedings of the Aristotelian Society* before the war.

51. Quoted in Gilson, *La philosophie au moyen âge*, p. 229.

52. Quoted in Ueberweg, p. 510.

53. Though Gilson's account of Scotus in *La philosophie au moyen âge*, pp. 225-242, assumes the genu-

.inity of certain works later held to be spurious, it would be premature to reject his interpretation as too severe. Cf. his article, "Les Seizes Premiers Theoremata et la Pensée de Duns Scot," in *Archives d'Histoire Doctrinale et Littéraire du Moyen Age*, 1937-1938, pp. 5-86, esp. pp. 46-80.

54. Quodl. II, q. 4: "Concludi potest quod non potest demonstrari Deum esse infinitae virtutis in vigore seu intensive, quia si posset demonstrari, non posset nisi per productionem alicuius de nihilo, scil. per creationem . . . sed per creationem demonstrari non potest, quia nec ipsa demonstrari potest." Quoted in Michalski, *op. cit.*, pp. 73-74, from FL 15372.

55. "Quia melius est tenere cum eo ad quod vadit intentio ecclesiae . . . pono propositionem . . . quod licet demonstrari non possit animam esse formam corporis . . . tamen tenendum est, secundum quod mihi videtur, quod anima est pura actuatio et formatio corporis." Quoted by Michalski, *op. cit.*, p. 75, from FL 15867, fol. 96-vo, col. 2.

56. *La philosophie au moyen âge*, p. 229.

57. The connection would be as follows. Matthew of Aquasparta, one of St. Bonaventure's pupils, was so alive to the difference between knowledge of universals and knowledge of singulars that he considered two different kinds of "species" to be needed for these two kinds of knowledge. Scotus' distinction between intuitive and abstractive

knowledge tends to emphasize the same difference, although not in the same way.

58. Scotus' use of this distinction to explain angelic knowledge is preceded by a clear explanation of the distinction itself. Cf. the text in the Vives edition, Paris, 1893 (juxta editionem Waddingi), Vol. XI, pp. 212-213.

59. On the *Summae Logicales*, Cf. Michalski, *op. cit.*, p. 62.

60. According to Michalski, *ibid.*, p. 76, in the first period of Paris nominalism "we find not a single decree of condemnation of this doctrine either from the University or from ecclesiastical authority; the decrees of condemnation appear only at the time when the Paris theologians, following the example of Ockham, introduced into their lectures the subtle distinctions between 'suppositio simplex' and 'suppositio personalis, propria et impropria' which we have found in the Venerabilis Inceptor of Oxford."

Ockham's "suppositio impropria" is thus described (Cf. original text in his *Summa Logicae*, as quoted in Michalski, *op. cit.*, p. 67, from Ms. Bibl. Jag. 719, fol. 26) :

"A term can always have personal supposition, no matter in what proposition it is used, unless it is taken in another supposition because of the will of the users. . . . Thus it often happens that genuine and authoritative (*magistrales*) propositions are false if taken literally *(de virtute sermonis),* but they are true in the sense in which they are put forward: and this sense is, what the speakers meant by these true propositions."

It is easy to see the effect which such a distinction would have on interpretation of the Scriptures. And in general, the meaning of any term would be quite independent of its accepted usage.

61. "Quam positionem sustinui in secundo meo principio contra magistrum G. Calcar, et nunc oppositum hic teneo, non quod magis unum putem verum quam aliud, sed ut coloretur multipliciter imaginandi via." Quoted by Michalski, *op. cit.,* p. 63, from FL (new) 1467, fol. 195-ro, col. 1.

62. "Ad articulum principalem nihil assentiendo sed tamen gratia exercitii disputando dico, quod Deus vult peccatum fieri et esse, et hominem peccare mortaliter et venialiter." Quoted by Michalski, *op. cit.,* p. 69, from FL 16400, fol. 96-ro, col. 2.

63. *La philosophie au moyen âge,* pp. 205, 208.

64. *Ibid.,* p. 206.

65. *Ibid.,* pp. 209-210, 217-218.

66. *Ibid.,* p. 210. For texts see Ueberweg, p. 470.

67. *Ibid.,* pp. 214-215.

68. Quoted in Gilson, *ibid.,* p. 215. For other texts Cf. Ueberweg, pp. 472-473.

69. *La philosophie au moyen âge,* p. 216. On Peter of Maricourt, Cf. text in Ueberweg, p. 465.

70. *La philosophie au moyen âge,* pp. 217-218.

71. Michalski, *op. cit.,* pp. 73-75. Also *La philosophie au moyen âge,* pp. 246, 248.

72. Abbé K. Michalski, *Les courants critiques et sceptiques dans la philosophie du XIVe siècle.* In *Bulletin International de l'Académie Polonaise des Sciences et des Lettres,* (Cracow, 1925, pp. 41-122) p. 97.

73. *La philosophie au moyen âge,* p. 251; *The Unity of Philosophical Experience,* pp. 78-82. This interpretation of Ockham, at least insofar as it finds in his views the seeds of skepticism, has been strenuously objected to by Philotheus Boehner, O. F. M., in an article which appeared since this lecture was delivered ("The Notitia Intuitiva of Non-Existents according to William Ockham," *Traditio,* Vol. I, 1943, pp. 223-275, esp. pp. 235, 238-240). In reviewing this article (Cf. book review section of *Thought,* June 1944), A. C. Pegis promises to discuss it fully in the second volume of *Traditio.* Without anticipating the results of this discussion, I can only say that a careful study of the texts quoted in the article and a careful re-reading of Gilson's comments have brought to light no reason why I should change anything in the text of this lecture.

74.   *Les courants philosophiques à Oxford et à Paris pendant le XIVe siècle*, pp. 78-80.

75.   *La philosophie au moyen âge*, pp. 278-279. The texts are referred to in *The Unity of Philosophical Experience*, pp. 97-102.

76.   *La philosophie au moyen âge*, pp. 269-273. Note the excellent six-point summary of his views on knowledge, p. 269.

77.   *Ibid.*, p. 216.

78.   Quoted in Gilson, *La philosophie au moyen âge*, p. 273; for source, Cf. Ueberweg, p. 593.

79.   Ueberweg, *loc. cit.*

80.   See *Science*, September 17, 1943, Vol. 98, p. 250.

81.   Quoted by Gilson, *La philosophie au moyen âge*, p. 278.

82.   That the Ockhamists alone were not responsible for this development is clear from Abbé K. Michalski, *La physique nouvelles et les différents courants philosophiques au XIVe siècle*. In *Bulletin Internationale de l'Académie Polonaise des Sciences et des Lettres*, (Cracow, 1927, pp. 93-164), esp. 158; Cf. also *Les courants critiques et sceptiques dans la philosophie du XIVe siècle*, pp. 41-122, esp. 115-116.

83.   Cf. Ueberweg, pp. 597-598.

84.   *La philosophie au moyen âge*, pp. 290-293.

Universitas
BIBLIOTHECA
Ottaviensis

La Bibliothèque
Université d'Ottawa
Echéance

Celui qui rapporte un volume après la dernière date timbrée ci-dessous devra payer une amende de cinq sous, plus un sou pour chaque jour de retard.

The Library
University of Ottawa
Date due

For failure to return a book on or before the last date stamped below there will be a fine of five cents, and an extra charge of one cent for each additional day.

24 OCT 1960

APR 2 4 1963

MAR 2 5 1964

DEC 1 0 1968

19 SEP. 1990

DEC 0 6 1990

DEC 2 0 1999

DEC 1 4 1999